What can you see and do at Kew...?

Useful plants

Palm House - page 2

Museum No 1 - page 10

Plant survival

Princess of Wales Conservatory - page 12

Alpine House - page 34

Climb and scramble amongst the **Treehouse Towers** - suitable for 3-11 year olds. Play in **Climbers and Creepers** if you are 9 years or younger.

Kew around the world

Temperate House - page 20

Evolution House - page 26

Marianne North Gallery - page 28

Pagoda & Japanese Gateway - page 35

Become an explorer

Xstrata Treetop Walkwa

and Rhizotron - page

Palm House

Inside the Palm House, it is like the steamy tropical rainforest.

Did you know?
The Palm House has 16,000 panes of glass. No wonder it takes weeks to clean them all!

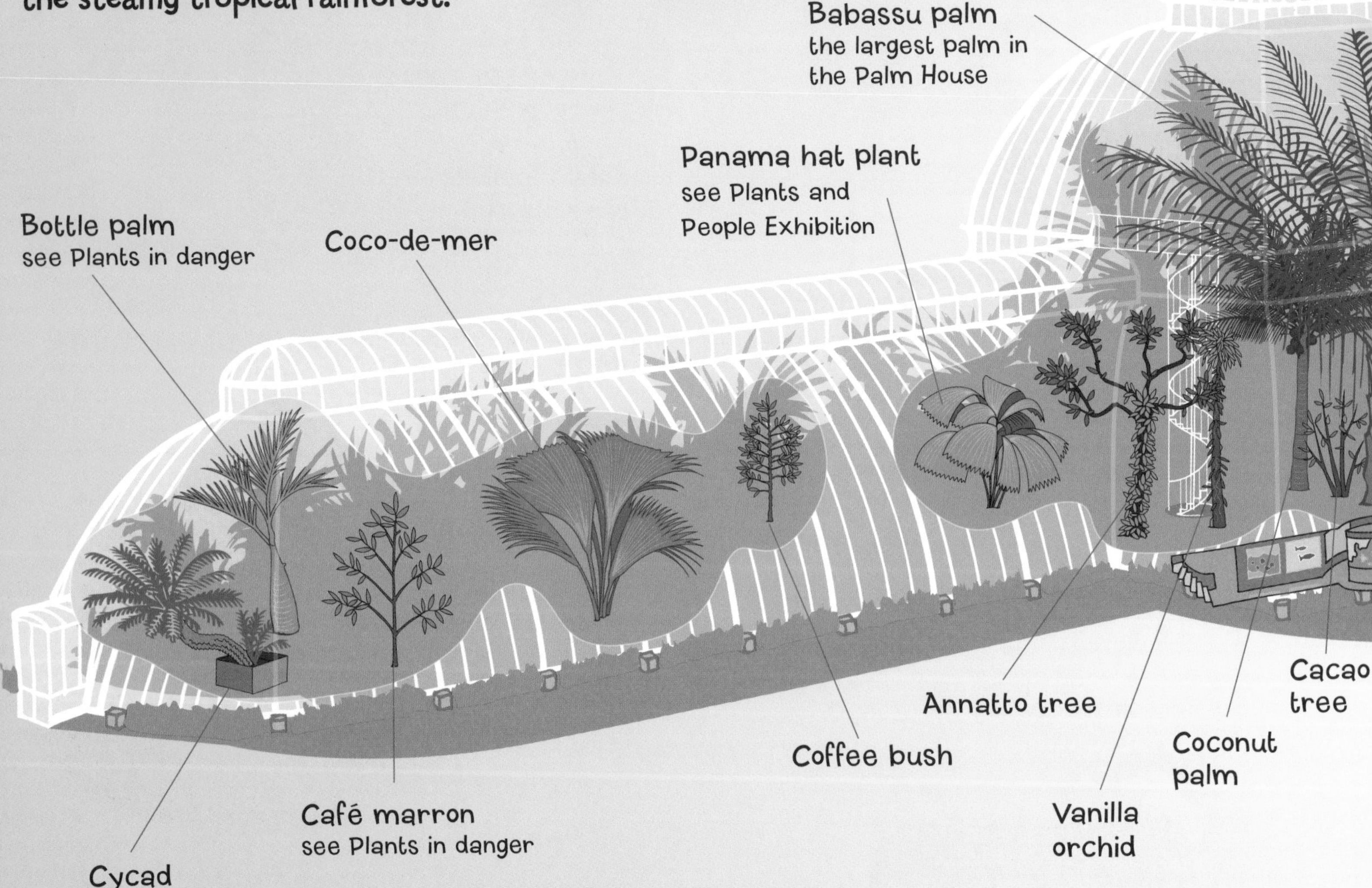

Put a sticker next to each plant on the list that you spot on your jungle jaunt.

Giant bamboo

Mango tree
only produces fruit in the tropics

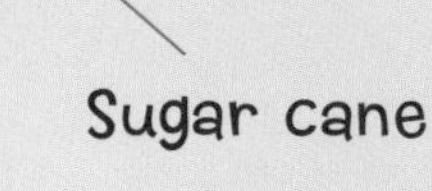

Sugar cane

Rattan palm

Banana
look for green unripe bananas

Marine Display

Rubber tree

Find the plants

Annatto tree

Rattan palm

Cacao tree

Rubber tree

Coffee bush

Coco-de-mer

Cycad

Sugar cane

Giant bamboo

Vanilla orchid

You can find out about these plants on the next few pages.

Inside the Palm House

Find rainforest plants that are ingredients for some of the world's favourite foods and drinks.

Chocolate, yum!

Cocoa beans grow inside pods that sprout from the trunk of the cacao tree. The beans are piled into warm, soggy heaps to bring out their delicious chocolate flavour. They are dried and shipped to chocolate factories.

Curries, cakes and doormats

Coconuts are the seeds of the coconut palm. The white kernel inside is used in many foods from curries to cakes. The coarse fibres on the outside of the coconut can be made into doormats.

Food colouring

The orange colour in many foods comes from the soft flesh that surrounds the seeds of the annatto tree. In tropical America, native people use the colour as body paint and lipstick.

Did you know?
A coconut can kill, if it falls on a person's head!

Ice-cream flavour

Vanilla flavour comes from the seed pods of this tropical climbing orchid. The pods have no flavour until they are sweated and dried. Natural vanilla is expensive so most vanilla flavouring is artificial.

Vanilla orchid

How many ants can you see on this page?
Answer at the back of the guide.

Wakey, wakey!

Most of the world's coffee comes from bushes grown in tropical America. There are two coffee beans inside each red fruit. The beans need to be roasted before they smell and taste of coffee.

Coffee bush

Fizzy drinks can have loads of sugar.

Sweet grass

Sugar cane is a tall grass, which grows in the tropics. The canes are shredded and crushed to release the sweet juice inside. Over two-thirds of the world's sugar is made from sugar cane.

Sugar cane

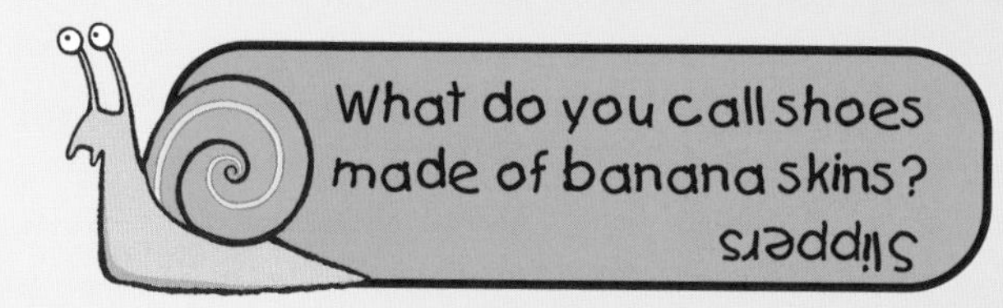

What do you call shoes made of banana skins?
Slippers

Inside the Palm House

Look for these rainforest plants, which help us to build and make things. You can also find some record-breakers here.

Cutting rattan

Cane chairs

Rattans are climbing palms with long stems that are made into cane chairs, baskets and mats. The palm has spines that help it to scramble high in the forest canopy. The spines may hurt collectors harvesting rattans from the rainforest.

Rattan

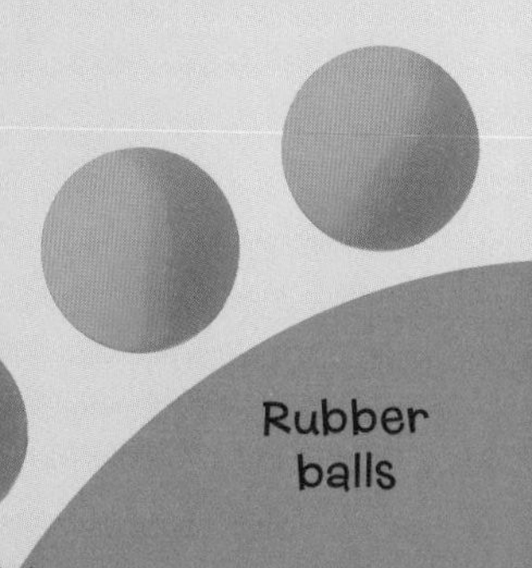

Rubber balls

A tapper collecting latex

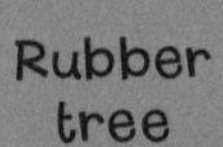

Rubber tree

Bouncy balls

White liquid, called **latex**, oozes from cuts made in the bark of the rubber tree. The latex is dried and made into stretchy rubber. Balls made of rubber bounce because they spring back into shape.

Coco-de-mer

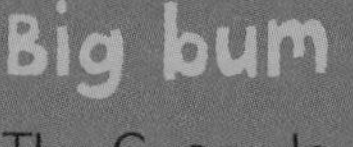

Coco-de-mer seed

Big bum

The Coco-de-mer palm has the world's largest seed, whi looks like a big bottom. The seed weighs up to 30 kilos. After a seed thuds to the ground, it takes over two ye before it starts to grow.

Really ancient

Cycads are a group of plants that thrived 200 million years ago, when dinosaurs liked to eat them! This cycad in the Palm House is over 200 years old.

The cycad is Kew's oldest pot plant.

Fast grass

Giant bamboos are the world's fastest growing plants. They grow up to 25 metres high, which is as tall as a stack of 16 cars. Bamboo is used for scaffolding in south-east Asia because it is strong and flexible.

Bamboo scaffolding

I'm a rainforest pitcher plant. Unscramble the words to find plants you might need if you got lost in the rainforest without supplies.

Clue — all the plants are on the Palm House map!

1. Make a shelter using alpm leaves and omobab poles.
2. Take cane from the tartan palm to make a chair.
3. Collect some tonococus, sanbnaa, asrgu neca, and esnoamg to eat.

Answers at the back of the guide

Palm House Marine Display

Head downstairs in the Palm House to discover plants that live on the shore and in the sea.

Drifters

Microscopic plants, called plankton, drift in the sea. They give us much of the oxygen we breathe. Plant plankton is food for small drifting animals, which are in turn eaten by fishes, and even the blue whale, the largest animal on Earth.

Remember to put on the glasses to see plankton in 3D!

Coral reef

Some tiny plants are partners with animals, called corals. They help the corals build huge reefs in tropical seas. Coral reefs are home to many kinds of fishes and other animals. Most reef fishes are brightly coloured to signal to others on the crowded reef.

Mangrove swamp

Mangrove swamps are found mostly in the tropics where fresh water mixes with salt water. The mangrove shrubs are partly submerged when the tide comes in. Prop roots anchor the plants in the mud.

British estuary

Estuaries are where rivers meet the sea. In cooler parts of the world, such as Britain, salt marshes are found in estuaries. Plants growing in the mud bind it together and help to protect the coast.

Rocky shore

When the tide goes out on a rocky shore, you can see different coloured seaweeds. Snorkel over the shore (in a safe place to swim) when the tide is in and the seaweeds look like a miniature forest.

Why is the sand wet?
Because the seaweed!

My snail cousins live at the seaside. Please colour in the seaweeds on their menu.

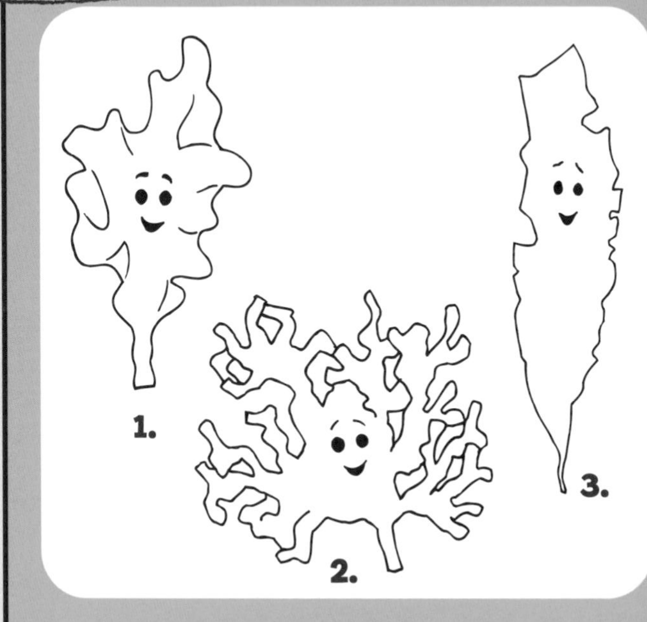

1. Sea lettuce – I'm green. I live where fresh water runs down the shore.
2. Irish moss – I'm red. I live in rock pools and low down the shore. I am eaten in bramble flan in Ireland.
3. Kelp – I'm brown. I live low down the shore and in deeper water. Jelly taken from my leafy fronds keeps your ice cream soft.

Plants and People Exhibition

Discover more useful plants and some beautiful and bizarre things made out of plants.

Cannibal Cutlery

This fork and plate from the Fijian islands are made of wood from the intsia tree. They were given to Kew in the mid 1800s when fortunately the practice of eating people was coming to an end.

You would not be here without plants! We give you oxygen, food, shelter, medicines, clothes, and more.

Panama hat

Panama hat plant

Cool hats

Hats made from dried leaves keep you cool in summer. The largest hat on display is made from leaves of the coconut palm. The panama hat shows how strips from the leaves of the hat plant are woven together.

Poisonous plants

People in the Amazon rainforest coat their arrows and blow-darts with different poisons. The poisonous roots of a vine are sometimes used for arrows. A drug made from this poison was also used to relax a patient's muscles during surgery.

Quiver for poison arrows

Heart pills

Chemicals from foxgloves have been used in heart medicines for over 200 years. You can find foxgloves growing in the Conservation Area in summer.

Papyrus scroll

Ancient scroll

The stalks of papyrus were made into a material to write on in Ancient Egypt. The papyrus plant grows in wet places in tropical Africa including along the river Nile.

Why did the tomato blush?

Because he saw the salad dressing.

You can see a coconut palm and panama hat plant in the Palm House.

Can you pick the correct answers to these questions about the Plants and People Exhibition?

1. What item of clothing on display is made from pineapple leaves?
   a. socks **b.** shirt **c.** pants **d.** jumper
2. What plant produces a seed case, which you can use to scrub your back?
   a. sponge **b.** bottle-brush plant **c.** prickly pear **d.** loofah
3. Half the world eats the seeds from which kind of grass?
   a. rice **b.** corn **c.** bamboo **d.** rye
4. Wood from which tree is used to make cricket bats?
   a. oak **b.** sycamore **c.** pine **d.** willow
5. What plant dyes your jeans blue?
   a. blackberry **b.** indigo **c.** blueberry **d.** strawberry
6. What poisonous seeds are made into jewellery?
   a. rosary **b.** deadly nightshade **c.** lupin
   d. golden rain tree
7. What part of plants belonging to the cucumber family can you play music with?
   a. leaves **b.** roots **c.** gourds **d.** flowers

Answers at the back of the guide.

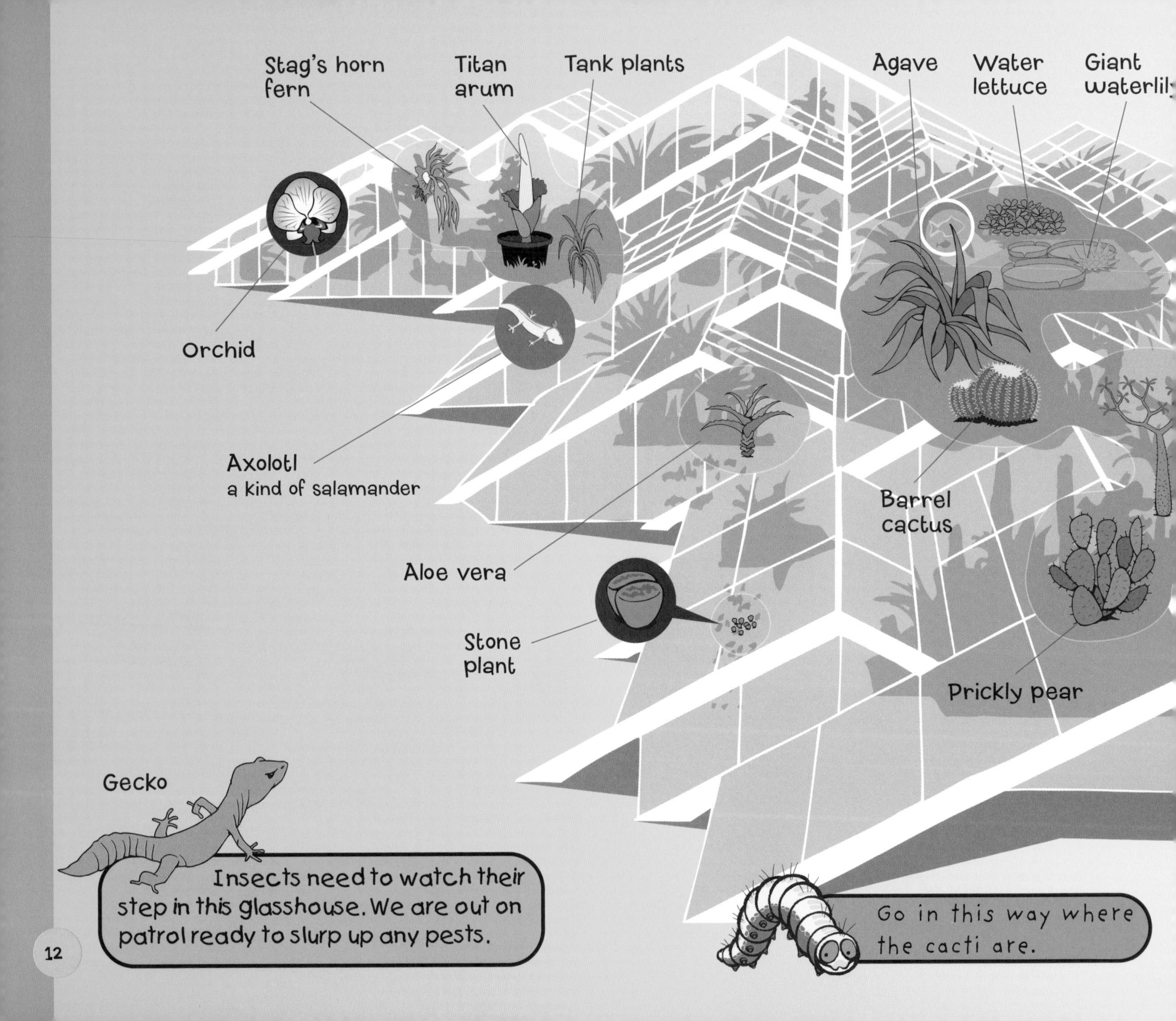
Stag's horn fern
Titan arum
Tank plants
Agave
Water lettuce
Giant waterlil
Orchid
Axolotl
a kind of salamander
Barrel cactus
Aloe vera
Stone plant
Prickly pear
Gecko
Insects need to watch their step in this glasshouse. We are out on patrol ready to slurp up any pests.
Go in this way where the cacti are.

Pitcher
plant
Sundew
Venus
flytrap
adagascan
ee

Princess of Wales Conservatory

Warm, wet or dry - which tropical climates do you prefer? You will find many climates in this amazing glasshouse.

Put a sticker next to each plant on the list that you spot on your walk around the desert and rainforest.

Find the plants

Agave

Orchid

Stone plant

Tank plants

Aloe vera

Pitcher plant

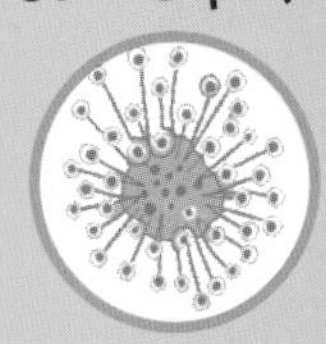
Sundew

Titan arum

Barrel cactus

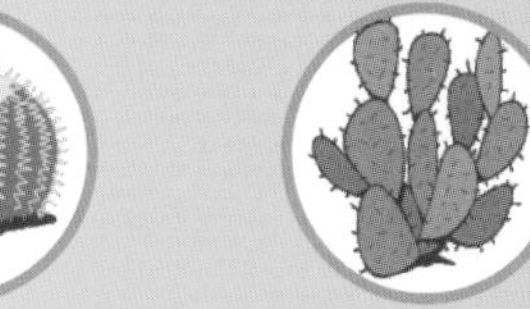
Prickly pear

Stag's horn fern

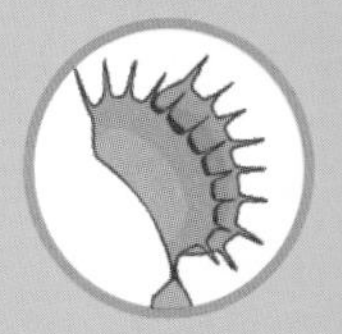
Venus flytrap

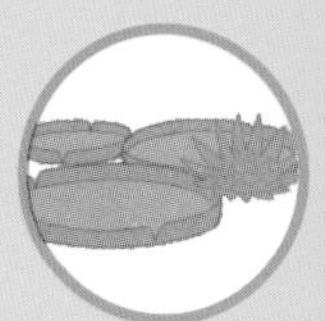
Giant waterlily

Madagascan tree

You can find out about these plants on the next few pages.

Inside the Princess of Wales Conservatory

Seek out plants that grow in deserts and other dry places.

Barrel cactus

Watery insides

The barrel cactus grows in American deserts, where it can survive for long periods of time on its store of water. The thick waxy outer layer keeps the water in and acts like sun-block.

Living stones

Stone plants look like stones, which may help them escape the notice of animals that might eat them. These plants grow in South Africa's drylands When stone plants flower after the rains, they are much easier to see.

Stone plants

Madagascan tree

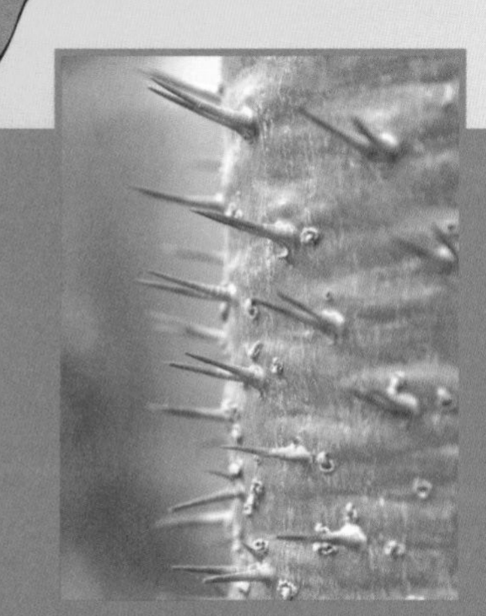

Madagascan tree spines

Spiny trunk

This unusual tree comes from the dry forests of Madagascar. The spines on the trunk collect moisture from dew. It drips onto the ground then is sucked up by the tree's shallow roots.

Peel me first

The fruits of the prickly pear are covered in sharp spines. By carefully peeling the fruits, people can eat them. The orange flesh inside is also made into sweets.

Prickly pear

Did you know?
Conservatory is another name for a glasshouse.

Lookalike

Agaves look rather like aloes but they come from the other side of the world. Agaves live in dry places in Mexico, south-western USA and Central and South America. Each circle or rosette of leaves only flowers once, then dies.

Agave

Say aloe

Aloes come from Africa and Arabia. They survive in dry places by storing water in their thick leaves. The jelly-like liquid from the leaves of aloe vera soothes the skin.

Aloe vera

Aloe vera lotion

Do you have shampoos or lotions made of aloe vera in your bathroom?

Fit the plant names into the puzzle.

Cactus
Stone plant
Prickly pear
Agave
Aloe

Answers at the back of the guide.

Inside the Princess of Wales Conservatory

Enter the lush wet zone where you may encounter giant leaves and flowers. Explore more to see orchids and ferns.

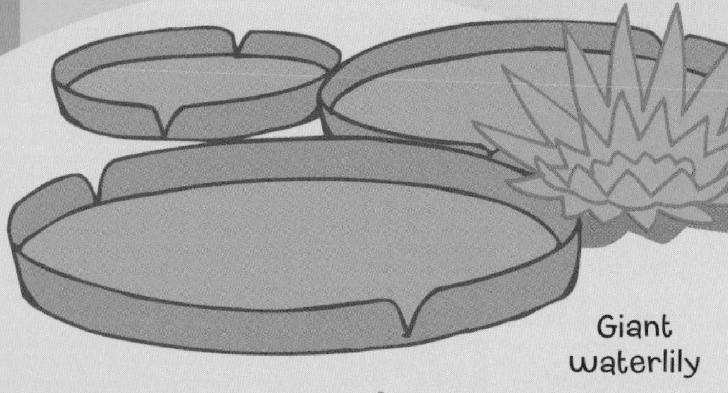

Giant waterlily

Giant waterlily ribs

Water lettuce

Lettuce menace

The water lettuce grows at the surface with its roots hanging down into the water. In some places, such as Florida in the USA, it is a menace. The dense growth blocks the light, and kills underwater plants.

Waterlily wonder

Giant waterlily leaves can grow over two metres across in summer. The leaves float because they trap pockets of air between the ribs underneath. The ribs are armed with spines to keep fish from nibbling at them.

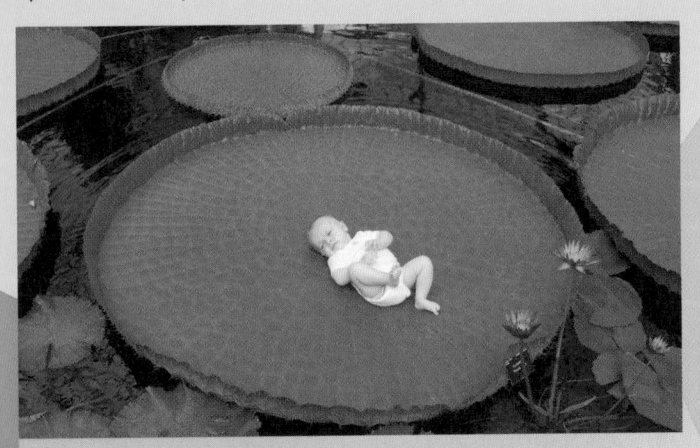

The floating leaf can take the weight of a baby. Please do not try this yourself!

The titan arum flower grows taller than a person.

A single giant leaf grows up afer the flower dies back.

Big stink

You may see the titan arum in bloom in summe It is the biggest flower in the world and smells foul. The flower grows u from a huge corm (a type of bulb). The titan arum comes from the Sumatran rainforest in Indonesia.

Bromeliads on trees

Perfect pools

Tank plants perch high up on rainforest trees. The leaves collect water to make a little mid-air pool in which frogs sometimes raise their tadpoles.

Tank plant or bromeliad

Showy orchids

Orchids have many beautiful flowers. Their seeds are tiny and blow away in the wind like dust. Most tropical orchids grow on tree branches, where their dangling roots catch moisture from the air.

Orchid on a tree

Tropical orchid

Fern waterfall

Fabulous ferns

The stag's horn fern grows on branches in the rainforests of south-east Asia. It is too high up to get nutrients from the soil. Instead, the fern collects debris in its leaves.

Stag's horn fern

Hello I'm an axolotl. Can you say my name 'ax-o-lot-al'? How many of my babies can you see in this pool? Come and see me in the Conservatory.

Answer at the back of the guide.

Inside the Princess of Wales Conservatory

Discover plants that digest animals, and go down the ramp to see who is underwater.

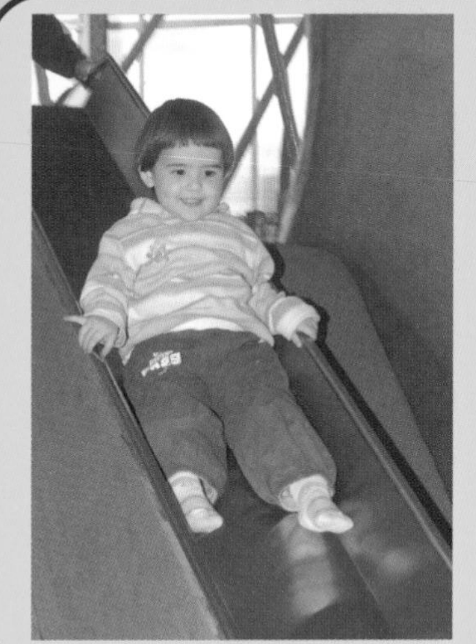

Have fun with us monsters in Climbers and Creepers.

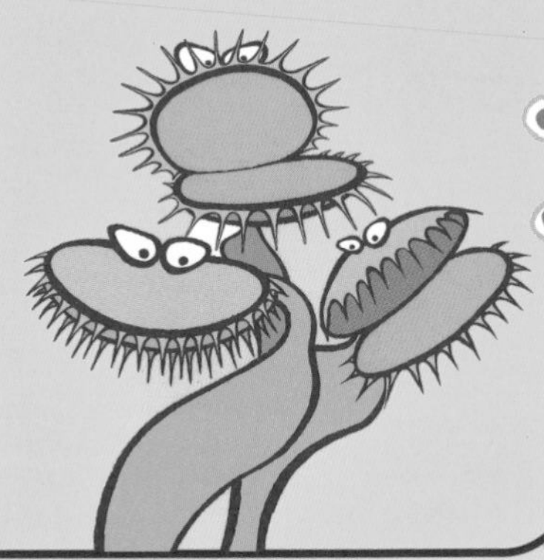

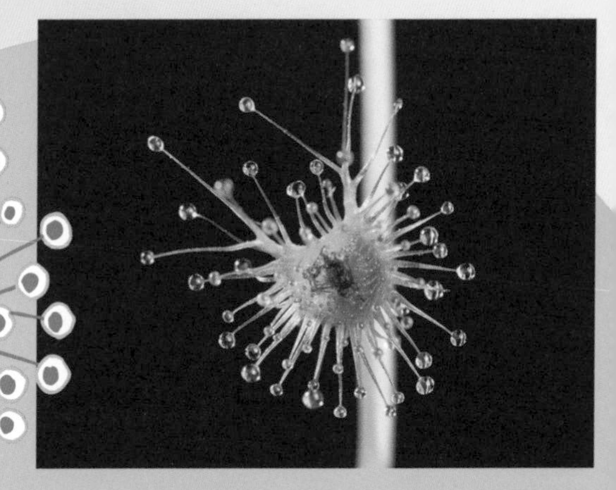

Sundew

Stickers

Sundews have droplets of glue to trap insects - just like flypaper. In some sundews, the leaf slowly wraps around the struggling victim to suffocate it. The dead insect is then digested.

What goes 99-bonk, 99-bonk, 99-bonk?

A centipede with a wooden leg.

Feed me

Carnivorous plants need to digest animals because they live in boggy places, where there are not enough nutrients in the soil. These meat-eating monsters have cunning tricks to trap small animals - most of which are insects.

Venus flytraps also digest spiders.

Snap trap

An insect feeds on nectar at the base of the teeth but when it brushes against trigger hairs several times the leaf snaps shut. The teeth interlock to prevent the insect escaping. The leaf squeezes the insect and starts to digest it when still alive.

Slip and slide

An insect is lured deep inside a pitcher plant by large helpings of nectar. The deeper it goes, the more slippery the wall becomes. The insect slips and tumbles into the liquid at the bottom where it drowns and is digested.

Small animals, such as frogs, sometimes fall into large pitchers and get digested too.

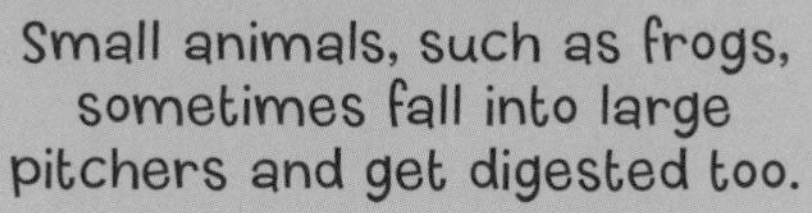

Pitcher plant

Piranha fish

Nasty nashers

You can see a piranha in one of the smaller tanks downstairs. The piranha has powerful jaws armed with razor-sharp teeth that rip out chunks of flesh. Fortunately, most kinds are shy of people.

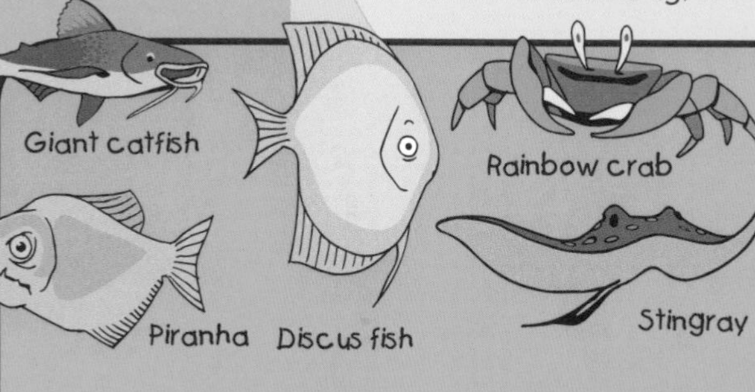

Water dragon

Can you find the names of our animals in the word search?

P	I	R	A	N	H	A	X	R	U	Q	A	S	O	V	T	W
C	K	O	B	Q	B	I	P	L	E	C	U	T	X	B	M	E
R	A	I	N	B	O	W	C	R	A	B	N	I	B	T	S	L
G	I	A	N	T	C	A	T	F	I	S	H	N	Q	F	I	V
Q	L	E	K	X	T	C	X	B	T	V	O	G	U	S	E	K
A	Q	D	R	M	O	W	A	T	E	R	D	R	A	G	O	N
K	E	X	D	N	Y	R	T	Z	W	Q	B	A	F	E	X	I
E	J	V	K	B	X	D	T	A	F	J	E	Y	T	H	M	D
G	X	D	I	S	C	U	S	F	I	S	H	E	I	D	C	N

Please note our animal displays change from time to time.

Answers at the back of the guide.

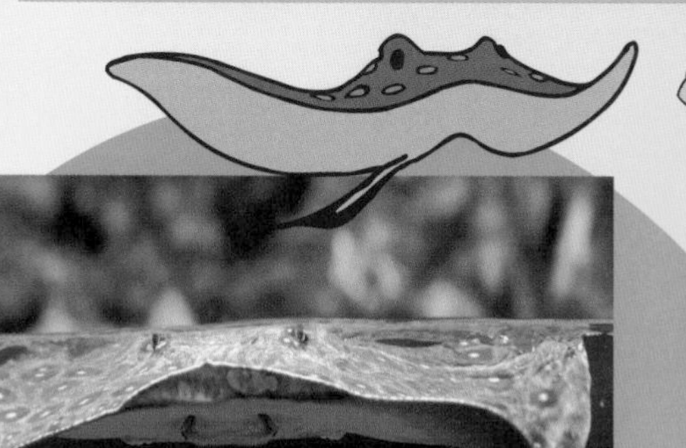

Freshwater stingray

Sting in the tail

f you look through the round tank windows downstairs, you can see freshwater stingrays. Their tails have sharp spines loaded with venom. Stingrays only attack when stepped on or harassed.

Temperate House

You can find plants here from the warm temperate zone that is between the frozen poles and steamy tropics.

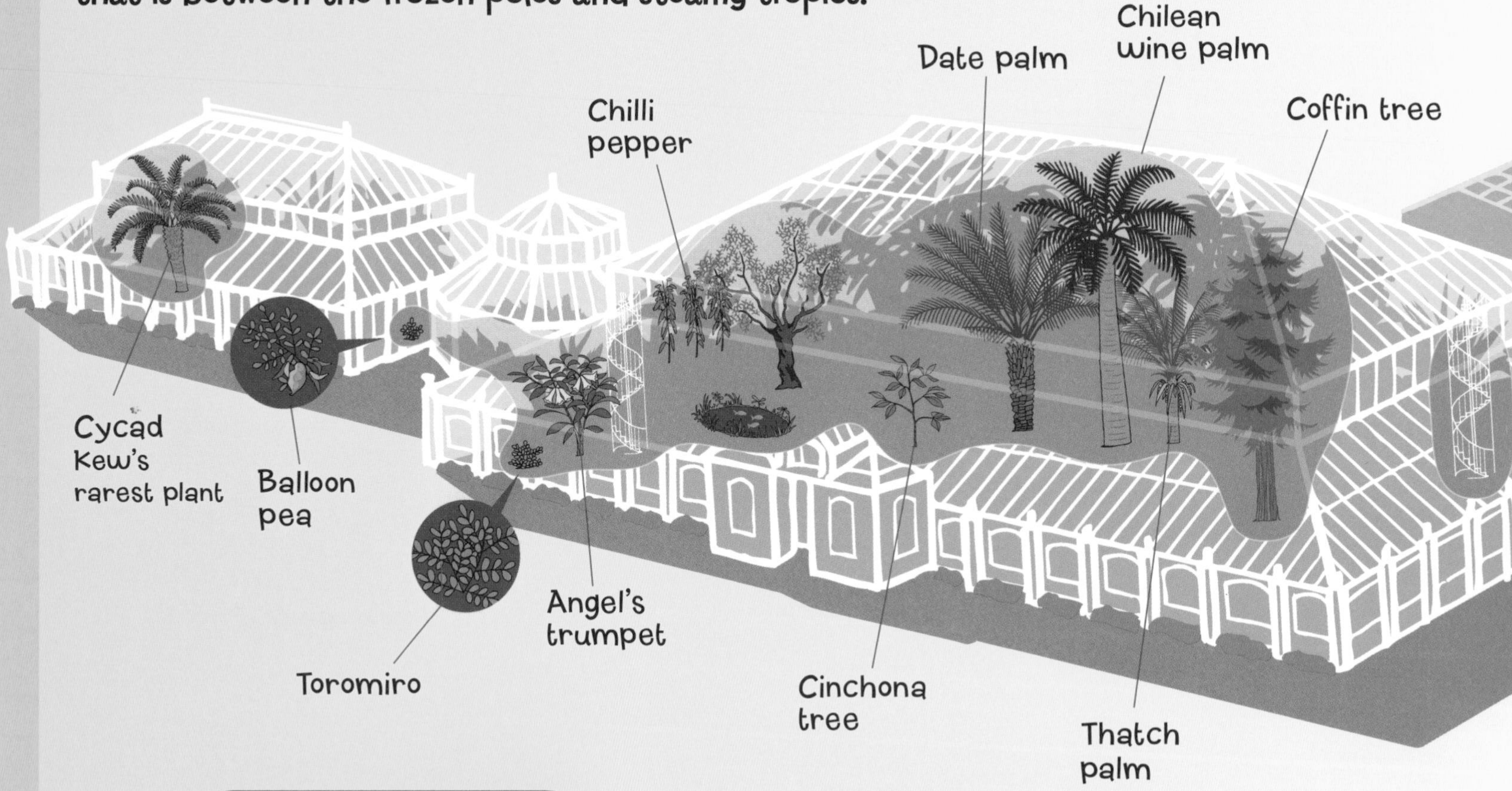

Knock, knock.
Who's there?
Leaf.
Leaf who?
Leaf me alone!

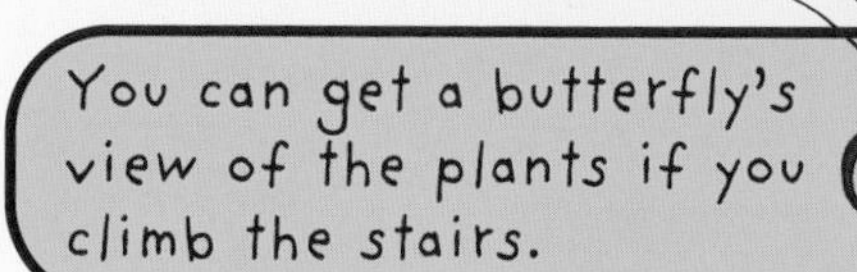

You can get a butterfly's view of the plants if you climb the stairs.

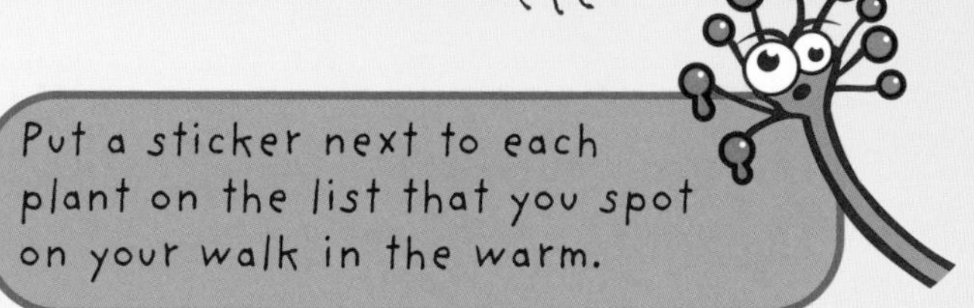

Put a sticker next to each plant on the list that you spot on your walk in the warm.

Tea bush

Find the plants

Angel's trumpet

Coffin tree

Balloon pea

Date palm

Chilean wine palm

Cycad

Chilli pepper

Tea bush

Cinchona tree

Thatch palm

You can find out about these plants on the next few pages.

Inside the Temperate House

Discover how plants from warm temperate places keep us happy and healthy.

You can see a toromiro tree in the Temperate House.

Wild no more

The last wild toromiro tree was chopped down for firewood in 1960 on Easter Island in the Pacific Ocean. New trees have grown from seeds planted at Kew.

Chilean wine palm

Tall tales

Kew's wine palm stands over 17 metres tall. It is the world's tallest indoor plant. Palm wine is made in Chile from the sugary sap that drains out of a palm after it is cut down.

Dates

Date palm

Yemeni hat

Make a date

The date palm has 800 different uses. Here are just a few of them! The leaves are made into hats, rope and thatch. The trunk is made into shutters and doors. The dates are made into sweets and cakes.

Cinchona tree

Life Saver

Bark of the cinchona tree produces quinine, which is a bitter-tasting drug. This is still used to treat malaria. The deadly disease (passed on by mosquitoes) kills up to three million people a year.

Out of Africa

Traditional remedies are important for healthcare in many parts of the world. In South Africa, the balloon pea is used as a tonic for people suffering from illnesses including AIDs and cancer.

Balloon pea or cancer bush

Join the dots to find out who is lurking in the pool.

Answer at the back of the guide

Inside the Temperate House

Spot plants from warm places with some strange uses. Keep a look out for Kew's rarest plant.

Chilli pepper

Red hot

Chillies are hot and spicy, which puts most animals off eating them. Many people like the taste. Birds cannot taste the fiery flavour. They eat the chillies and disperse their seeds when they poo.

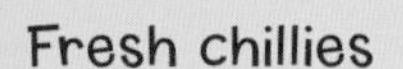

Fresh chillies

Coffin tree

The coffin tree comes from Taiwan in Asia. The strong wood is made into coffins for grand funerals. Coffin trees grow as tall as 65 metres, taller than Kew's Pagoda.

Chinese coffin

Coffin tree

Thatch or kentia palm

Home and away

The thatch palm comes from Lord Howe Island off the east coast of Australia. It has been grown as a potted palm since Victorian times (mid 1800s). Most potted thatch palms you see have come from seedlings raised on this island.

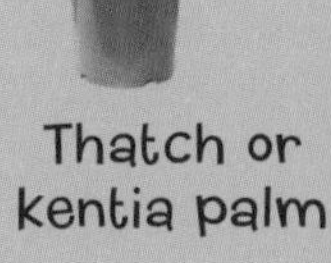

Thatch or kentia palm

Lonely boy

This male cycad grew from a stem collected from a solitary plant in South Africa over 100 years ago. No other plants were ever found. Today, this cycad only survives in gardens like Kew.

The male cycad has a cone but needs a female to make seeds.

Angel's trumpet flowers

Angel's trumpet

Buried alive

Angel's trumpet comes from Colombia in South America. In ancient times, the slaves and wives of dead Mayan kings were put to sleep by drinking a potion containing angel's trumpet, and then buried alive.

Hello, I'm a sundew.
Can you help me spot five different things in the Temperate House? The white box is a sticky trap like me that catches insect pests.

Answers at the back of the guide

Evolution House

Step back three-and-a-half billion years to begin your journey. Spot giant creepy crawlies along the way.

Early land plant *Cooksonia*

Liverworts

Pillow-like structures found in Australia today.

Way back

Three-and-a-half billion years ago, the only plants on the planet are single cells. These plants live in the sea. They form mats of gooey green slime. Some mats are covered in layers of sand to form pillow-like structures.

Leap forward

Around 420 million years ago, there are small plants on land for the first time. Some of these become extinct, such as *Cooksonia*. Simple plants, called liverworts, are still with us today.

Tree ferns

Clubmoss and giant dragonfly

Can you find the small mosses on the rocks and the giant millipede?

Step into the swamp

Around 320 million years ago, giant clubmosses tower into the sky. Some are as tall as Kew's Pagoda. Ferns and horsetails grow much taller than they do today in the huge swamps.

Look for horset[…] with their ribb[…] stems.

Did you know?
You can dig for fossils in Climbers and Creepers.

Fossil fuel

As the giant clubmosses and other plants die, they sink into the swamp. Millions of years later, their remains have become fossilized and turned into coal.

Coal

You can find many conifer trees in the Gardens today.

Jump into the Jurassic

Around 200 million years ago, conifers, cycads and other plants that produce seeds are doing well. Dinosaurs feast on ferns, horsetails and cycads.

Cycads have cones and make seeds.

What kind of dinosaur slept all day?

The dino-snore.

You can find butterflies in the gardens and in Climbers and Creepers.

Magnolia flower

Flowers flourish

Around 120 million years ago the first flowers bloom. They are pale and sweet-scented to attract beetles to carry their pollen. Showy colourful flowers only begin to bloom when there are bees, butterflies, and birds on the planet to pollinate them.

Triceratops

Walk on to today

Walk in a dinosaur's footprints. Some dinosaurs eat the soft, juicy flowering plants. Flowers are all around us today, but dinosaurs go extinct 65 million years ago.

Plant explorers

For over 200 years people from Kew have travelled the world in search of plants.

Beginning with Banks

Sir Joseph Banks (1743-1820) was a friend of George III and an intrepid explorer. Banks sent the first plant hunters from Kew on expeditions to bring back rare and useful plants.

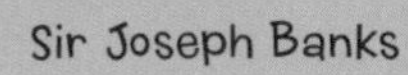

Sir Joseph Banks

Sir Joseph Dalton Hooker

Rhododendrons

In the Himalayas

Hunting for new kinds of rhododendrons in the Himalayas was gruelling for Sir Joseph Dalton Hooker (1817-1911), who became head of Kew. He was infested with 100 blood-sucking leeches after tramping the foothills.

The marvellous Miss North

Marianne North (1830-1890) was a plant explorer and artist. She came from a rich, well-connected family. When she was young she met lots of her father's friends, including Sir William Hooker, then Director of Kew Gardens.

Travelling the world

For 14 years Marianne travelled to many countries including Japan, India and Australia. In the days before widespread travel and photography her landscape paintings gave people a rare glimpse of these far-away places.

Painting places and plants

Marianne worked hard on her travels. Amazingly, she produced over 800 oil paintings. These show all kinds of plants growing in fields, forests, on mountains, by rivers, near the sea. Many are of plants close up, showing their flowers and fruits.

Discovering unknown species

Some plants Marianne painted were unknown to the experts at Kew and elsewhere. A few of these were later named after her. This one is called *Nepenthes northiana*.

If you like paintings, why not visit the Marianne North Gallery, and also colour in her flower painting on page 39.

Become an explorer

Take a trek around the Gardens to track down some terrific trees.

Scramble up the Xstrata Treetop Walkway for an exciting view. Discover what lives around tree roots in the Rhizotron.

Xstrata Treetop Walkway

Climb 118 steps to walk amongst the treetops. Every season brings new views of our trees. High above the ground, you can see far into the distance.

California giants

The coast redwood is the world's tallest tree towering over 115 metres. The giant redwood is the biggest with a trunk measuring 31 metres around. Redwoods take hundreds of years to reach record sizes. Our trees are youngsters with the oldest planted in 1912.

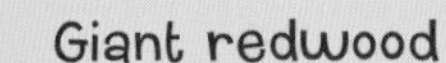
Giant redwood

This Japanese pagoda tree actually comes from China!

Prop me up

Nearly 250 years old, this pagoda tree is one of our oldest trees. Its lopsided look is probably because it grew with a double trunk - one half of which snapped away about 100 years ago.

Big puzzle

The monkey puzzle tree comes from Chile and Argentina in South America. The tree got its name when first grown in England because a person joked it would be hard for a monkey to climb.

Monkey puzzle tree planted in 1846.

The oak's leaf has tooth-like edges like a sweet chestnut leaf.

Chestnut-leaved oak

What a whopper!

The chestnut-leaved oak is Kew's biggest tree. It was planted in 1846. The tree grows over 30 metres tall and the branches spread 30 metres across, which is nearly three times the length of a London bus.

How many trees grow at Kew?
a. 700 **b.** 5000 **c.** 8900 **d.** 14,000
Answer at back of the guide.

Wollemi pine planted in 2005.

Still living

Scientists thought the Wollemi pine became extinct two million years ago. In 1994 pines were discovered living in a gorge near Sydney, Australia. Kew is helping to grow more Wollemi pines because there are so few in the wild.

Unlucky tree

A small plane flew into the top of this tree in the early 1900s. Since then it has been struck by lightning twice. You can see the split in the bark left after it was struck in 1992.

Corsican pine planted 1814.

Ginkgo

The female tree sprouts seeds that smell awful when they rot. If you pick up the seeds from the ground, their sickly smelling goo may irritate your skin. A stinky female tree is marked on the map.

Ginkgo leaves

Smelly ginkgo seeds

Become an explorer

Watch out for wildlife up in the trees, in the water and hidden under rotting logs.

Spot squirrels and jays carrying off acorns to hide. Look for pine cones gnawed by squirrels to get at the seeds inside. Find fantastic fungi (mushrooms and toadstools) but remember some kinds are poisonous.

Jay

Squirrel

Poisonous fungi

Wildlife friendly

We like to encourage wildlife at Kew. You can watch birds in the hide in the Conservation Area and join in a pond dipping session at the Dipping Pond.

Wildlife watch

Visit the ponds and lakes to find these baby wild water birds.

Coot with young

Mallard ducklings

I go out looking for food when everyone has gone home to bed.

Why did the mushroom go to the party?

Because he was a fun guy to be with.

Meet the tree gang

The tree gang are expert at climbing trees. They use ropes and pulleys like mountain climbers. They have to saw off branches that might otherwise fall on your head.

Stag beetle heaven

The Loggery is made of rotting logs, home to the grubs of Britain's largest beetle, the stag beetle. The grubs spend over four years feasting on the rotting wood before turning into adults.

Stag beetle sculpture in the Loggery

Take a wildlife walk in the Conservation Area

How many of my mini-beast friends are living in the log pile?

Woodlouse

Millipede

Centipede

Ground beetle

Earwig

Stag beetle grub

Answers at the back of the guide.

Kew does not use peat because digging for peat destroys the boggy places where sundews like me grow.

Buildings old and new

Visit some beautiful old buildings. Witness the latest in glasshouse design at Kew.

Kew Palace

This royal palace was once the home of George III (1760-1830). He spent time here as a child in the school room, and also when he fell ill with a disease that made him seem mad.

These buildings are on the map of the Gardens. Use the coloured boxes on this page to help you find them.

Keep us cool

Warm air is drawn out of the top of the Alpine House so that the mountain plants inside do not overheat. Cool air is blown onto the plants from an underground maze. This works in the same way as the cooling passages in a termites' nest.

The Davies Alpine House opened in 2006.

Queen Charlotte's cottage

Did you know Kew was home to tigers as part of the Royal Family's collection of animals in the early 1700s?

Royal cottage

The thatched cottage was given to Queen Charlotte when she married George III. The Royal Family had picnics here. Kangaroos and exotic birds were once kept nearby.

Bombs away

This Chinese-style pagoda was built in 1762 for George III's mother. At that time it had 80 golden dragons perched on the roofs. In World War II bomb designers dropped dummy bombs from the top floor to test them.

The Pagoda

We are Japanese

Our Japanese Gateway is a much smaller copy of one in an ancient city in Japan. Can you see the gateway's beautifully carved flowers and animals? Stroll around the gateway to see what gardens look like in Japan.

Minka

Recycled house

The Japanese Minka is a house made of wood. It has a thatched roof and mud plastered walls. This traditional style house was taken apart in Japan and rebuilt here. Minkas withstand earthquakes, which are common in Japan.

Amazing paintings

See flower paintings old and new at the Marianne North Gallery and the Shirley Sherwood Gallery of Botanical Art.

Why do ducks watch the news?

To get the feather forecast.

Plants in danger

More and more plants are at risk as forests are chopped down, and wild places are taken over by towns and farms.

Time for trees

Much of the dry forest on the coast of Peru in South America has been cut down for farming and fuel wood. We are working with local people to plant more huarango (say wa-ran-go) trees.

Children help to plant trees in Peru.

Our world may be getting warmer because of people's energy-guzzling lifestyles. Plants like me may die out.

Seed bank saviours

Over a billion seeds from around the world are kept safe at Kew's Millennium Seed Bank at Wakehurst Place. We are collecting and storing seeds from places where plants are at risk, and we are helping other countries to set up their own seed banks.

By 2020 we hope to have a quarter of the world's different kinds of wild plants saved in the Seed Bank.

Jungle adventures

Today, Kew helps people around the world to look after their plants. Expeditions to the rainforest in Cameroon, West Africa aim to discover which plants there are under threat.

Wild food

Kew's scientists are looking at traditional foods in Africa, such as roasted seeds from the sausage tree. Traditional foods may help people in Africa stay healthy and find enough to eat in times of famine.

The sausage tree is named for the shape of its fruit - not because it grows sausages!

Bottle palm

This palm only grows wild on Round Island, Mauritius in the Indian Ocean. In the 1970s, it almost died out because rabbits and goats like to eat it. Seedlings from Kew have been planted back in the wild.

You can see the bottle palm in the Palm House.

Some of our rarest plants are grown in the laboratory.

Which plant lives where?

Match the plant to where it lives

Harebell

Desert

Rocky shore

Stag's horn fern

Grassy meadow

Cactus

Rainforest

Seaweed

Answers at the back of the guide.

Colouring fun

Colour in this dragon from the Japanese Gateway. Can you find a dragon in front of the Palm House?

Follow in Marianne North's footsteps and colour in her painting of the sacred lotus from Java. Go to page 28 to read about her adventures.

Puzzling fun

What do I carry from flower to flower so that they can make seeds? Fill in the puzzle to see the answer in the shaded squares.

1. When are tulips in bloom?
2. What kind of bulb is used with cheese to flavour crisps?
3. What part of a plant is green and traps sunlight?
4. What do bees do to get around quickly?
5. What chocolate treats do you eat in spring?
6. What is the name of the sweet substance in flowers?

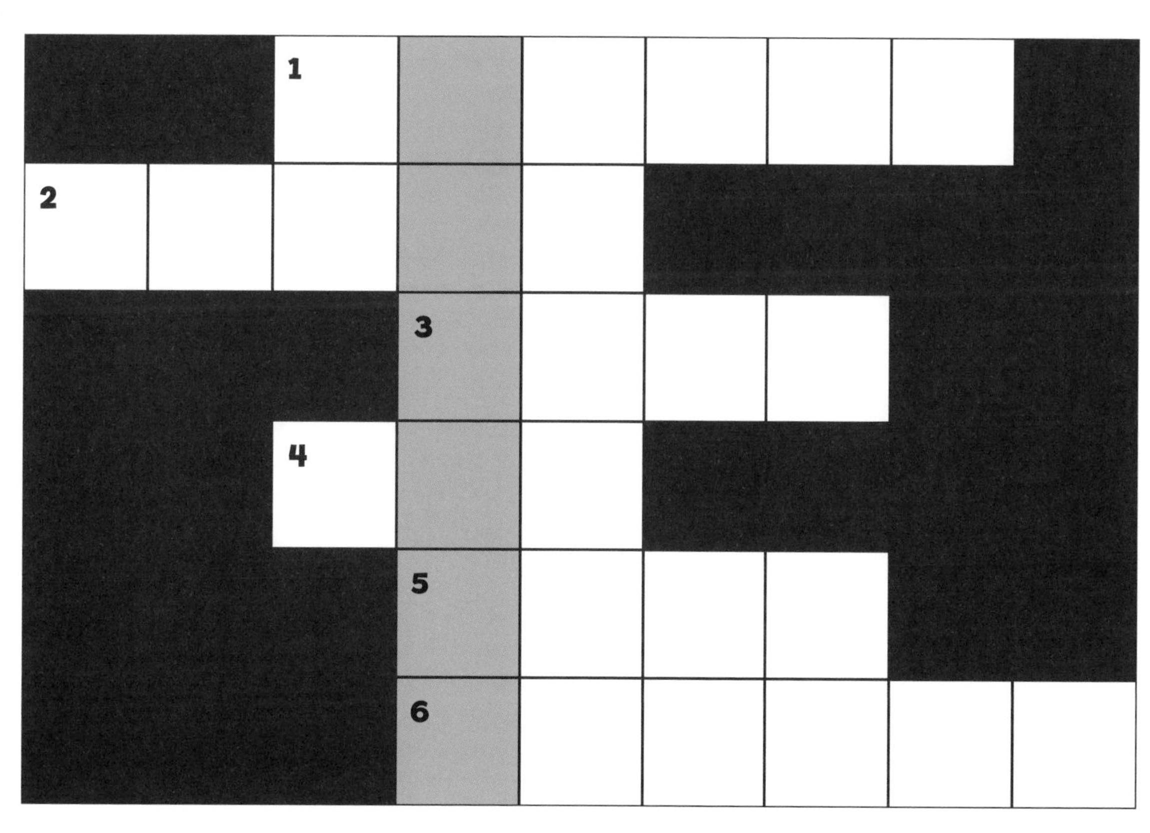

Answers at the back of the guide

What does a caterpillar do on New Year's Day? Turns over a new leaf.

Activity fun

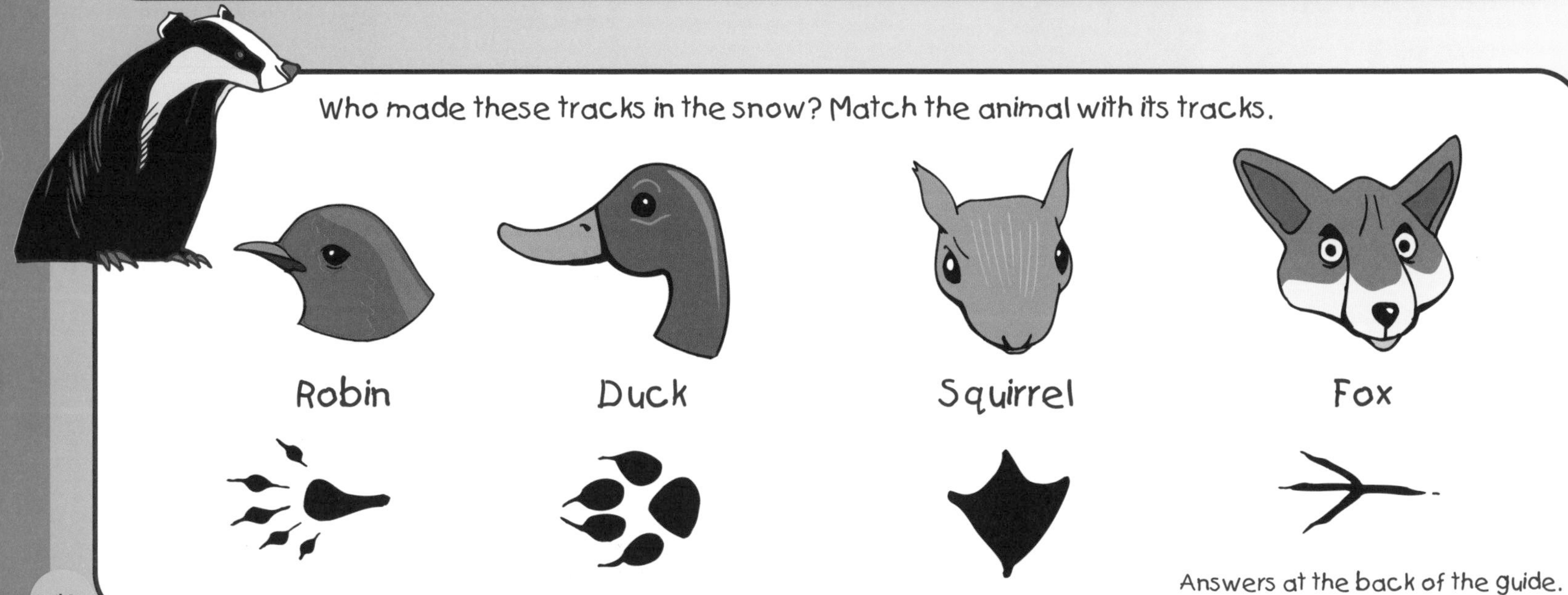

Grow a hyacinth

Wear gloves if you have sensitive skin. Balance a hyacinth bulb on top of a narrow jam jar full of water. The roots will grow to fill the jar. The bud opens into a scented flower.

Grow a Sunflower

If you want tall sunflowers in your garden or container in summer, you can give them a head start by planting seeds in small pots indoors in spring.

Plant an oak tree

Pick up an acorn and plant it in a pot outside.

Your baby oak should sprout in spring. When it out-grows the pot, you can put it in a larger container.

Make a bark rubbing

Bark is the outer layer of woody plants including trees. The colour and pattern of the bark can help you identify a tree. To make a record of the pattern take a rubbing of the bark.

You will need:

- **strong paper**
- **wax crayons or chalk**

Place your paper on the bark. Rub the crayon or chalk lengthwise across the paper.

Bark rubbing

Make a pine cone animal

Olly Owl

Use your imagination to make cone animals such as birds, mice, etc.

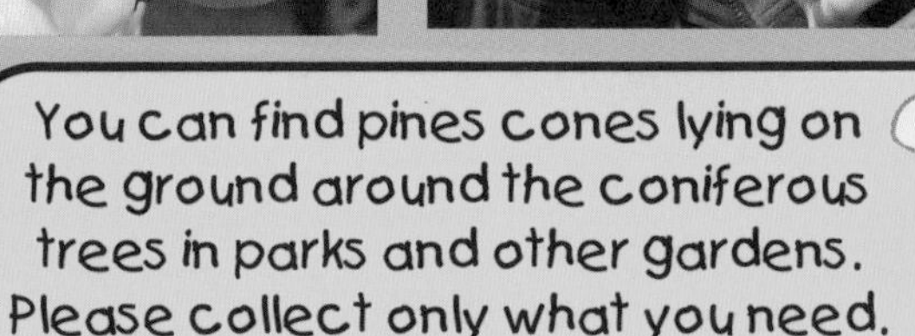

Dinosaur garden

You will need:

- a washing-up bowl
- a bag of gravel
- attractive stones
- peat-free potting compost
- bottle-garden plants (you can buy them in garden centres)
- toy dinosaurs

Put 5—6 cm of gravel in the bottom of the bowl. Then fill it to within 1 cm of the brim with potting compost.

Create a cave, valley or cliff and prehistoric landscape with the stones, then position the plants. When you're satisfied with how it looks, plant your little bottle-garden plants.

Release your dinosaurs, letting them, and your imagination, roam.

Pressed flowers

Have fun pressing garden flowers.

You will need:

- flowers from your garden
- some old newspapers
- several sheets of kitchen or blotting paper
- a stack of heavy books

Pick flowers when they are dry (no rain or dew). Pansies and poached-egg plants press well. With fat flowers like roses and poppies, it is better to press individual petals.

Place an old newspaper on a table and put kitchen or blotting paper on top. Arrange your flowers on the paper. Cover with blotting or kitchen paper. Top off the pile with another newspaper and a stack of heavy books. Leave your flowers to dry for two weeks.

Use pressed flowers to make cards or pictures, or stick them on to paper to study. Botanists still dry plants to study like this. A collection of pressed plants is a herbarium – Kew's Herbarium is very big.

REMEMBER Always ask permission to collect flowers or weeds. Don't collect wildflowers, as many are becoming rare.

Here are some things to do at home

I'm a harebell – I live in wild grassy places. You can help wildflowers like me and other wildlife by doing some of these things.

Dig a pond.
Pile up old logs.

Build a bird table and a bird bath.
Put up a bird or bat box.

Grow wild flowers to help bumblebees.

In dry summers use dirty bathwater to water plants.

Leave grass and nettles to grow in a sunny place in your garden as food for caterpillars.

Use a water-butt to collect rainwater for watering your plants.

Answers Page

Pages 4 and 5
Number of ants on the pages = 32.

Page 7 Word scramble
Make a shelter using **palm** leaves and **bamboo** poles.
Take cane from the **rattan** palm to make a chair.
Collect some **coconuts, bananas, sugar cane** and **mangoes** to eat.

Page 11 Plants and people quiz
1b. shirt, 2d. loofah, 3a. rice, 4d. willow, 5b. indigo, 6a. rosary, 7c. gourds

Page 15 Word puzzle

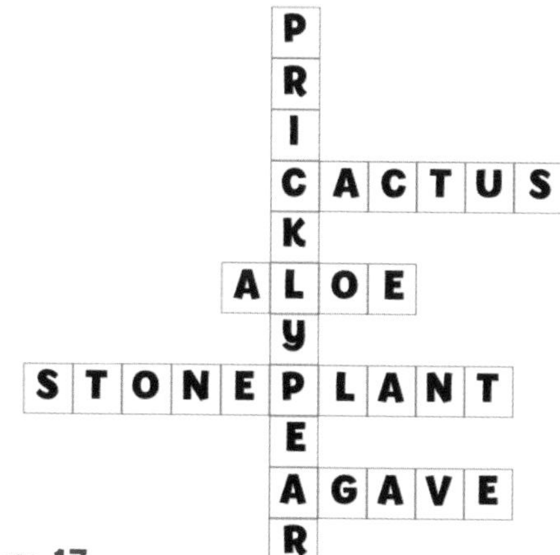

Page 17
Number of baby axolotls in pool = 5
The other 8 were adults!

Page 19 Word search

```
P I R A N H A X R U Q A S O V T W
C K O B Q B I P L E C U T X B M E
R A I N B O W C R A B N I B T S L
G I A N T C A T F I S H N Q F I V
Q L E K X T C X B T V O G U S E K
A Q D R M O W A T E R D R A G O N
K E X D N Y R T Z W Q B A F E X I
E J V K B X D T A F J E Y T H M D
G X D I S C U S F I S H E I D C N
```

Page 23 Who's lurking in the pool?
Koi carp

Page 25 Spot the difference
mushrooms, pool of water, leaf beside bucket, leaves in bucket, leaf on tree

Page 31 Number of trees
d. 14,000

Page 33 Numbers of mini-beasts in the log pile
Woodlouse = 4
Millipede = 3
Centipede = 3
Ground beetle = 4
Earwig = 3
Stag beetle grub = 3

Page 37 Matching game
Harebell lives in **grassy meadow**.
Stag's horn fern lives in **rainforest**.
Cactus lives in **desert**.
Seaweed lives on **rocky shore**

Page 41 Word puzzle

Page 42 Matching game

My name is Sticker the sundew.

This second edition published in 2011
by the Royal Botanic Gardens, Kew,
Richmond, Surrey, TW9 3AB, UK

Written by Dr Miranda MacQuitty
Illustrated by Guy Allen
Designed by Louise Millar
Development editor: Lydia White

ISBN 978 1 84246 431 1

10 9 8 7 6 5 4 3 2 1

www.kew.org

Printed in China by Midas Printing (within a factory that has ISO 14001 accreditation, the internationally recognised standard of environmental responsibility). The paper used in this book contains material sourced from responsibly managed and sustainable commercial forests, certified in accordance with the FSC (Forestry Stewardship Council).

The following people have provided invaluable help and advice:
Raffat Bari, Shirley Beadle, Sandra Bell, Fiona Bradley, Gail Bromley, Karen Brown, Henry and Rose de Chazal, Colin Clubbe, Lucy Cole, Simon Cole, Louise Cross, Ambar and Hannah Driscoll, Gina Fullerlove, Laura Geary, Phil Griffiths, Christina Harrison, Tina Houlton, Tony Kirkham, Lloyd Kirton, Jane, Emily and Michael Lambert, Paul Little, John Lonsdale, James and Oliver Morley, Mark Nesbitt, Christine Newton, Jill Preston, Sue Seddon, Varsha, Shivani and Ria Sokal, Katie Steel, Nigel Taylor, Amber Waite, Catherine Welsby, Lydia White.

We would like to thank the following for providing photographs and for permission to reproduce copyright material:
Acro Images/Alamy 19; Heather Angel/Natural Visions 8; Jeff Collett/ Natural Visions 8; Emma Dodd 43, 45; Redmond Durrell/Alamy 19; Jeff Eden 32; Peter Gasson 17, 32; Hemera Technologies/Alamy 11; Laura Jennings 17; Christabel King 37; Paul Little cover, 1, 5, 8, 10, 22, 23, 27; William Milliken 37; James Morley cover, 16; Norman T Nicoll/Natural Visions 8; RBGE/Ian Edwards 7; Brian Rogers/Natural Visions 26; StockStill/Alamy 32; Oliver Whaley 36; Lydia White 14, 15; all other photographs by Andrew McRobb.

For information or to purchase all Kew titles please visit www.kewbooks.com or email publishing@kew.org

Kew's mission is to inspire and deliver science-based plant conservation worldwide, enhancing the quality of life.

Kew receives half of its running costs from Government through the Department for Environment, Food and Rural Affairs (Defra). All other funding needed to support Kew's vital work comes from members, foundations, donors and commercial activities including book sales.

When you find a plant at Kew put a sticker on the plant lists in the guide.
Corsican pine
Rubber tree
Titan arum
Rattan palm
Cycad
Palm House
Giant waterlily
Stag's horn fern
Chilli pepper
Thatch palm
Tea bush
Monkey puzzle
Balloon pea
Venus flytrap
Pitcher plant
Giant bamboo
Aloe vera
Stone plant
Gingko
Agave
Coffin tree
Cinchona tree

Japanese pagoda tree

Annatto tree

Giant redwood

Tank plants

Can you find the stickers for the trees and other plants you spot in the Gardens?

Cycad
Temperate House

Cacao tree

Barrel cactus

Vanilla orchid

Madagascan tree

Coco-de-mer

Orchid

Angel's trumpet

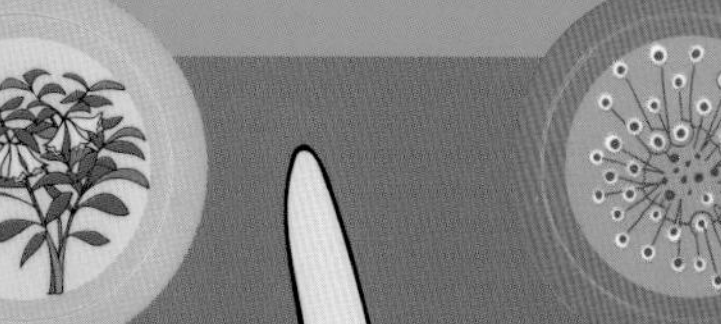

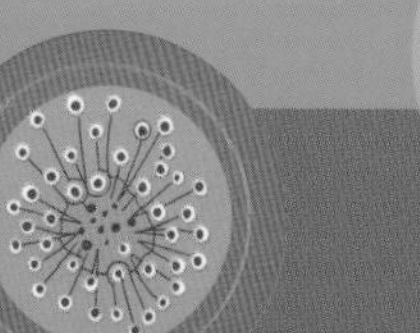

Sundew

Chestnut-leaved oak

Date palm

Chilean wine palm

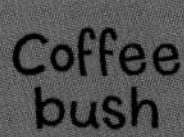

Coffee bush

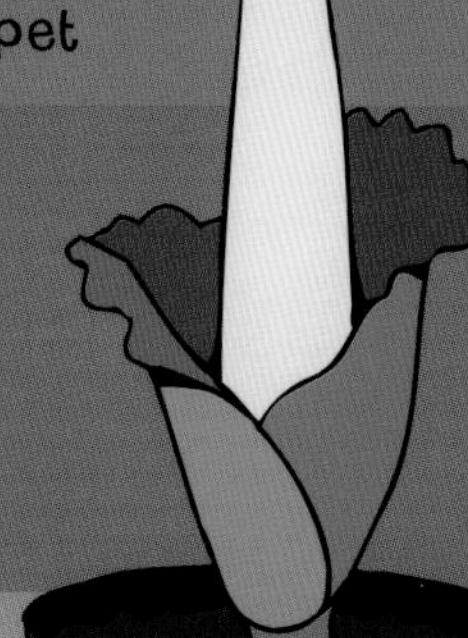

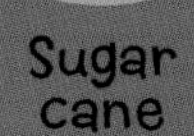

Sugar cane

Can you find the stickers for plants you spot in the glasshouses?

Prickly pear